Theory of Nothing

$$\sum_{t=0}^{t=\infty} 0 + \prod_{t=0}^{t=\infty} 0 \rightarrow \infty 0$$

Why Life is Unexplainable

Eric Scheuneman, MSc

First Edition - March 2008

ISBN 978-1-4357-1438-0

Published by Lulu.com

www.lulu.com

Acknowledgements

Thank You:

Teachers involved in my education
from kindergarten to university graduate school.

Federal Government managers
who saw beyond the surface of stutter to hire me.

Suzanne Scheuneman and Frans Vandendries
for support with editorial and conceptual comments.

Martha Scheuneman Holmes
whose enthusiastic reading of this book
appreciated and validated its spirit.

I continue to be blessed with family, friends and community.

I salute three pillars of Canadian society:

teachers, nurses and single parents.

Table of Contents

Foreword

This book and its theory are presented in a minimal manner and avoid exhaustive background material - which is in keeping with the true and profound simplicity of the theory itself. Readers will have the creative pleasure of thinking of their own examples and how they fit into the *Theory of Nothing*.

The equation on the cover and title page is taking poetic licence with the mathematical notation to symbolize that:

the *sum of all random independent creation processes*
of the Universe
from nothing (0) over infinite time (t)

$$\sum_{t=0}^{t=\infty} 0$$

plus (+)

the *product of all sequential dependent creation processes*
of the Universe
from nothing (0) over infinite time (t)

$$\prod_{t=0}^{t=\infty} 0$$

leads to (→)
infinite forms of nothing.

$$\infty 0$$

The *infinite forms of nothing*
are all the things that comprise the Universe.

Chapter 1

Thought and Explanation

We think, therefore we explain.

Why People Are Compelled to Fit Things into a System or Pattern

Once the human brain evolved to a certain level of consciousness, people realized two things:

(1) thinking as power to explain and control things, and

(2) mortality and its implications.

It is likely that these two realizations are the driving force for most people's quests in life.

The nature of these two realizations is very powerful and primal and became paramount, even surpassing sex and survival, once the human brain crossed the threshold of conscious intelligence.

For many people, this longing for the certainty of explanation of life or for the meaning and implications of mortality is a deeply embedded need resulting in many people feeling a void in their lives which leads them to searching for something to fill this void.

Many people spend much time - some a lifetime - trying to fit things into a system or pattern. This effort spans all areas of life: religion, science, medicine, economics, sports, and so on. In later chapters we will look at various ways people try to fill this void.

However, first we will have a brief look at the ultimate expression of thinking power: the quest to develop a theory to explain the laws that govern the entire Universe.

Unified Field Theory

Albert Einstein, in 1905, conceptualized the most famous equation in the modern world:

$$E=mc^2$$

which is so simple and elegant. He spent the last years of his life trying - and failing - to develop a unified field theory (1950) to encompass both the macro (gravitation) and micro (electromagnetism) worlds of physics.

Theory of Everything

After Einstein's time, we have had much experimental and theoretical work done by many brilliant scientists using the incredible power of computers to develop what has become known as the *Theory of Everything*. This has included the amazing physicist and popular author, Stephen Hawking.

A series of scientists developed *String Theory* which uses complex 'string' particles to attempt to explain such things as the apparent wave-particle duality of the atomic and sub-atomic world.

The latest exciting and beautiful development is by Garrett Lisi and titled *An Exceptionally Simple Theory of Everything*. However, the title is belied by the 29 pages of fairly high level mathematics of E_8 Lie algebra along with several assumptions or parameter uncertainties.

Attempts to reconcile the macro and micro in the Universe have become increasingly complicated and draw upon quite far-out concepts. The next chapter discusses why such efforts are so complex - and doomed to failure.

Chapter 2

Theory of Nothing

Why life is unexplainable.

Creation of the Universe

The creation of the Universe and life started from - *nothing* - the chaos or complete disorder of infinite space and formless matter which existed before the Universe came into being.

The formless matter was infinite, began at infinite density, and exists for infinite time. Hence, the property of being infinite - without limits or bounds - is inherent to space, matter, and time.

The combination of chaotic origin and being infinite confers the property of changeability: space, matter, and time can change. This change can be gradual or instantaneous or spontaneous or a combination over time and by circumstance.

The Big Bang said to all the infinite, formless matter:

Go forth into infinite space and forever multiply in like manner as the chaos of disorder that produced you!

And this has resulted in many things produced throughout the Universe, each with their own disorder - or order.

Theory of Nothing

1. Everything in the Universe has resulted and developed from the chaos of nothing through many different paths and processes over long periods of time.

2. To expect everything to fit together or be explainable is unrealistic and a denial of their origins.

3. The Universe is unexplainable - there is no overall system or pattern except for the chaos or disorder of the Universe - although some things may be explainable.

4. Space, matter and time are infinite and changeable.

The exact nature of the creation and development of the Universe is not critical to the foregoing theory which is a process that works from any starting point. Even principle 4 of the theory is not critical to the other principles and overall prediction of the theory.

Elaboration

Chaos or complete disorder existed and ruled at the beginning: so it should continue to rule through its creations.

This is supported by the fundamental thermodynamic law of entropy which states that the tendency of the Universe is toward increasing disorder. All observable facts and events of the Universe support this law - disorder rules.

Even though there are many observable things - living or inanimate - or facts or events of the Universe, there is no way to explain, order or relate them all under one theory or law.

The existence of pockets of order in the Universe is itself an affirmation of the *Theory of Nothing* because total disorder would be an impossible result from infinite paths and processes. Indeed, this is illustrated by the well-known infinite monkey theorem that a monkey typing at random for infinity will almost surely type the complete works of Shakespeare.

There are ordered and explainable things in the Universe which have occurred from random chance, direct cause and effect, indirect cause and effect - or a combination thereof. These are three main ways that ordered things can develop from a disordered Universe.

Scientists are still looking for the *Theory of Everything* - which is a contradiction of the principles of the *Theory of Nothing*. They are going to extraordinary lengths and complexities to try to make an overall theory work for everything. If the Universe formed quickly and cataclysmically and then developed over long periods of time by different events and circumstances, why should there be an overall theory or law?

Since scientists are trying to cover all the things resulting from the chaos of disorder, it is no wonder that they have had to turn to increasingly complex mathematical and physical constructs!

It has been said that simplicity is elegant and best for theories and laws. Chaos is the zenith of simplicity and elegance. It allows divergence and development through the chance of disorder.

The chaos of disorder and chance from *nothing* produces diversity which leads to everything which prevents explaining everything. In other words, the origin and development of the Universe leads to everything which - ironically and paradoxically - precludes a *Theory of Everything* and leaves us with the *Theory of Nothing*.

Chapter 3

Life on Earth

Life is a collection of interesting chaotic results.

Every thing in the Universe operates under its own principles or laws which developed simultaneously with the thing - and continues to develop and change - often gradually but sometimes quickly. Continual and differential development is another factor making full explanation of everything impossible.

Some things can be, more or less, fully explained. And there may be commonalities amongst different things: common laws or physical or chemical properties. However, in many cases the striving need to explain things results in extrapolation and over-reaching of single or a few facts or events to yield flawed theories or explanations - as examples will show later.

Life on Earth is a collection of things of the Universe that are unexplainable because of the complex chaotic creation and development of the Universe. Creation of life on Earth - governed by the overall Universe law of disorder - is part of the tendency of the Universe toward increasing disorder.

However, some aspects or parts of life have order or explanation. Having some order in some things is a natural consequence of chance and disorder.

Different laws governing different things are a natural outcome of chaos and chance since they represent disorder amongst the whole of the Universe. Laws on Earth - different from other parts of the Universe - are also part of this.

The following chapters will look at the various aspects of life on Earth to illustrate that life is a manifestation of chaos; life has developed from nothing in accordance with the process of chaos. Which entities are explainable, inter-related or unexplainable?

Each chapter will offer a few examples without being exhaustive, leaving the Readers to think of their own examples to analyse under the *Theory of Nothing*.

But first we need to take a look at what constitutes explainable.

Chapter 4

Explainable

Power to predict and change.

A thing is *explainable* when:

- it is clear and understandable such that:
  - its meaning is known;
  - its action and use is known;
  - its organization and formation is known.

For our broader purpose - which is why explainable is such a powerful driving force - we will add to this the implicit and inherent promise of explainable:

- its behaviour is predictable and changeable.

There is a certain irony - or duality - in predictable and changeable being paired together!

This relates directly back to the start of Chapter 1 - *power to explain and control things*.

If some thing is changeable by us, then it likely is changeable on its own as part of the on-going Universe development process that was proposed in Chapter 2. Of course, we - and our actions - are part of the Universe development process.

The reality is that some things will be partially explainable since they will not fulfill all the properties listed above. Fortunately, we often do not need to explain everything about a thing to get useful results!

Indeed, this is the kernel of why the *Theory of Nothing* is a positive result from the negative fact of not being able to explain everything.

Explainable offers the possibility of the power to change a thing and to predict observable events or facts concerning it; these are very useful properties beyond the intellectual satisfaction of basic understanding.

Although clarity and specificity are often desirable, we must be careful not to become distracted or paralysed by definitions and grey areas of classification of phenomena. Indeed, requiring exactness for everything in our Universe of chaos and disorder and change is contrary to the nature of the Universe - and leads to failure, such as the *Theory of Everything*.

There are different levels of structure or classification that we deal with when trying to explain some thing. What level do we want to deal with? For discussion and analysis it is useful to break some thing down into its components.

For example, the human body is a good example of some thing which has many components such as the circulatory system, various organs, immune systems and so on. Although it seems unlikely that we shall be able to explain the complete body or even all its components, it is still extremely useful to explain as many of its components as possible.

Another example would be a molecule which is comprised of atoms which are comprised of various sub-atomic particles. Learning about the components is useful for physics and chemistry which can be useful for biochemistry and the human body.

Going in the opposite direction, sometimes it is very useful for discussion and analysis to group things together. For example, different plants or animals can be grouped together into classes or species.

Both component and grouping approaches can be very helpful in trying to explain things in a usable manner as will be seen in the genetics chapter.

Power to Predict and Change

How do we achieve the power to predict and change some thing? Which components or groupings are explainable, inter-related or unexplainable?

A functional and useful approach to analyse some thing for explainable information may be diverse and even imprecise - like the predictions and results of the *Theory of Nothing* itself.

Sometimes the key will stem from only one of the explainable factors: meaning, action and use, or organization and formation.

Sometimes it will be serendipity or observation or trial and error.

Realizing that ordered and explainable things in life have occurred from random chance, direct cause and effect, indirect cause and effect - or a combination thereof - means that the clues to explainable can be found almost anywhere!

So, it is important to keep one's mind open to all the possibilities!

What are the key factors that indicate some thing is explainable? And explainable to what degree and use? We will look at this in the context of entities comprising Life on Earth which is of primary interest and is relatively accessible for observable facts and events.

The following chapters will illustrate various ways that the human mind has tried to make life explainable - tried to predict and change things.

Chapter 5

Religion

Religion is a construct of the mind to deny mortality.

Religions or gods - at various levels of sophistication - probably appeared before science because they are more primal, require a lower level of intelligence, and more easily and quickly satisfy the dread of mortality.

Religion or god is the one stone that simultaneously kills the two birds of mortality and explanation. It grants life after death and offers explanations for itself and life.

Why was food scarce? Why did children die? Why did people become ill? Why did eclipses occur? What happens after we die?

These would be some of the earliest questions arising in the increasing consciousness of the human mind. Since these questions were - and sometimes still are - unanswerable and unexplainable by the human mind, people likely turned to religion first to seek relief and protection from mortality and lack of understanding.

Initial religions likely focussed on the most obvious celestial bodies such as the sun and moon. Or they could focus on earthly features such as powerful animals or impressive landmarks. This was an intuitive association of power with understanding: trying to achieve understanding through association with a powerful thing.

As the mind and civilizations became more sophisticated, religions also became more sophisticated. They developed and incorporated many mythologies based on both real life and fantasy, including the god figures assuming human forms.

Soon politics would enter the religion field as individuals realized the power of being a seer or religious leader. Once human religious figures appeared they were naturally followed by explanations and pronouncements since humans expected information because they could directly communicate with the religious figures - in contrast to distant or inanimate gods. Positive and negative incentives were developed by religious leaders to encourage people to join and continue to follow.

Diversity of religions is a manifestation of the chaos disorder paths governing the Universe; like everything else, there were - and are - many ways for religion to develop. Further, the fact that there is no one answer - no *Theory of Everything* - means there is no one religion.

In the previous chapter, the power, satisfaction and value in being able to explain components was outlined. Diversity of religions also developed because various religions could satisfy certain groups of people at certain times. This is one way that paths are determined - and changed.

Religion could not exhibit the explainable property; it did not have the power to predict and change everything. Hence, faith developed as the answer or antidote to this failure. *Faith is the saviour of religion;* it is both a hope and a conceit to explain or satisfy everything.

Many people develop a fervour which likely stems from their sub-conscious longing for the certainty of explanation or meaning of life. This sub-conscious need results in many people feeling a void or dissatisfaction in their life which compels them to search for something.

For those who reject conventional religions for whatever reason, alternate religions or cults or philosophies have developed to satisfy these groups of humanity. For those who reject religion and its alternates, there are philosophies such as existentialism and nihilism - or science and humanity and the arts.

Chapter 6

Mathematics & Physics

We think, therefore we count.

Physics is the mind striving to explain matter and energy.

Mathematics is the scientific study of number, quantity, and space - either abstract or applied to real phenomena.

Physics is the scientific study of the nature and properties of matter and energy.

Physics and mathematics are inextricably connected since mathematics is the tool used to quantify and explain physics.

Chapter 1 presented - the concept of the ultimate expression of thinking power - physics and mathematics trying to develop a *Theory of Everything* to explain the law(s) that govern the entire Universe. In Chapter 2 the *Theory of Nothing* was proposed to show why an all-encompassing theory cannot exist. A few more comments will be added here.

Why should the life on Earth and in the Universe follow the same rules and laws? Each developed in their own ways and their own different paths. The attempt to reconcile the scientific macro and micro into the same mathematical laws is an extension of the concept in Chapter 1 that thinking should be able to explain and control everything.

Science is both a hope and a conceit that the mind should be able to explain everything in a systematic manner.

Science is the converse of religion: science (thereby man) understands and controls everything compared to god understands and controls everything (since man cannot).

Mathematics has developed many different types of formulas and regimes: some explain certain situations and some explain other situations. This is a manifestation of diversification - many branches or paths - resulting from overall disorder throughout the Universe developing from the chaos origin of life.

Physics began by using concrete or real mathematics to describe matter and energy. As scientists expanded their horizons and knowledge, physicists realized the limitations of their initial theories and models . Hence, physicists began to expand their horizons and techniques to incorporate abstract mathematics to try to explain matter and energy.

Since the later years of Einstein's life, physicists have been trying to develop a "Theory of Everything". Why have the world's greatest scientific minds with incredible computer tools failed?

Because everything in the Universe developed along so many different paths through so many different processes, mathematicians and physicists trying to explain the Universe have had to resort to increasingly complicated models to encompass this incredible diversity. These models are conceptually beyond anybody's comprehension. Even if a model appears to 'fit' what we appear to know, how can we use it when we cannot comprehend it?

There have been many problems even trying to understand and reconcile measurements and observations from only the macro Universe of outer and deep space. This could be due to different or changing laws governing different areas and times of the Universe which fit with the possibilities under the *Theory of Nothing*.

Some of the physics theories look attractive because they appear to match or explain certain or numerous features of observed physical phenomena. However, this is a result from the chaos disorder creation going in many directions and following many paths; some components are bound to correspond to some physics theories.

Lisi's E_8 Lie algebra theory might well offer an accurate description of the physical phenomena and some calculative results. However, the E_8 Lie group has 248 dimensions - compared to our normal perceptions of the 4 dimensions of length, width, height, and time. This shows to what extremes physics and mathematics must go in attempting to explain everything. This is understandable under the *Theory of Nothing* which states that nothing has - and still is - developing along infinite paths and processes to produce everything. Hence, Lisi's E_8 Lie algebra theory will not extend beyond the limits of its vision; it will not provide ways to explain and control things in everyday life.

Although there is great beauty and satisfaction in offering a complex description, the useful path is to develop what works for the various components of life.

Einstein's $E = mc^2$ is the best example of the elegance of simplicity yielding extremely useful practical results and benefits - while simultaneously showing the dark destructive side of knowledge. Even though Einstein was frustrated in not being able to develop an all-encompassing theory, we should realize - and utilize - the benefits of partial success.

There are several physical laws that appear to apply on Earth and might even apply throughout the Universe. However, they are not directly related or closely tied together. These include Einstein's equation relating energy to mass and the first and second laws of thermodynamics dealing with conservation of energy and entropy.

The existence of these laws with no clear relation between them is a result of the *Theory of Nothing*; they originated and developed wholly or partially independently of each other - and were 'discovered' by the minds and experiments of scientists.

Chapter 7

Biology & Evolution

Biology is the mind striving to explain living matter.

Plants and animals are the primary entities classified as living matter on Earth. They have been studied to various degrees since the human mind reached beyond the human body.

The existing recorded history of biology begins with the Greek philosopher Aristotle (4th Century BC) and has developed through many people in many cultures including Islamic physicians (8th - 16th Centuries AD) followed by European scientists such as Linnaeus (18th Century AD) and Darwin (19th Century AD).

Charles Darwin proposed and provided scientific evidence that all species of life have evolved over time from one or a few common ancestors through the process of natural selection. His theory of evolution has been generally accepted and modified somewhat.

The *Theory of Nothing's* first principle - *everything has evolved from the chaos of nothing through different paths and processes over long periods of time* - totally encompasses evolution theory. Further, it explains why there are many discrepancies in evolution theory and why evolution is not a neat and full picture and does not provide overall power to predict and change.

There cannot be a full continuous picture of the

development of life due to the infinite paths, processes and time involved. Also, it is unlikely that there was one common ancestor to life on Earth because of the many possibilities at every point in time. How does one even define the start? What preceded a certain starting point?

Even if there were one common ancestor to current human life on Earth, the myriad ways of development through such a long period of time probably render it un-trackable and unprovable. Genetics does an amazing job of correlating current genes, and an admirable job tracing backwards. However, genetics is constrained by limits of time, process, and possibilities resulting from the *Theory of Nothing's* principles.

A ground-breaking study - published in the science journal *Nature* in 2008 - supports and illustrates the *Theory of Nothing*: chaos over time precludes understanding and prediction of change in systems of life. Dutch scientists led by Jef Huisman carried out an 8-year study involving ocean life from the Baltic Sea which resulted in such conclusions as:

- shows a "chaotic" balance of nature;
- this chaos makes it impossible to predict the rise and fall of wild species - anywhere, ever;
- could never figure out a pattern that allowed real predictions of how any species would fare;
- advanced mathematical techniques proved the indisputable presence of chaos in this food web;
- short-term prediction is possible, but long-term prediction is not.
- it's like trying to forecast the long-term weather: all their forecasts broke down beyond a few weeks.

Further aspects of biology will be addressed in following chapters such as medicine and genetics.

Chapter 8

Medicine

Medicine strives to delay mortality.

Doctors - by necessity of helping people - are the perfect practitioners of analysing and working on components since conventional medicine has not been able to integrate life or the human body into one encompassing theory or practice.

Different parts or systems of the human or animal body may be explained - but there are no overall explainable connections. Even if there were, this would still neither translate nor apply to the Universe outside of the body.

This is understandable from the *Theory of Nothing* because the human or animal body developed from many sources over a long period of time according to its own methods and paths of disorder.

The need for explanation is very acute and strong in the field of medicine and illness. The consequences of ignorance and error can be debilitating or deadly. Hence, there has been much frustration and hope driving medicine - in both right and wrong directions.

Vaccination or Immunization

Vaccination or inoculation is the introduction, by injection or ingestion, of a substance into the body to produce immunity to a disease. The substance used in the vaccine can be live, weakened or killed bacteria or viruses - or purified proteins from such.

Inoculation against smallpox appears to have been carried out in India or China before 200 BC. It was known and used in Turkey in the 1600's. This was re-discovered by Edward Jenner in England from his and others' observation that dairy workers who had contracted the mild cowpox did not contract the deadly smallpox. Jenner's use of cowpox in 1796 to protect against smallpox was the beginning of modern vaccination. In the late 1800's Louis Pasteur in France built upon the work of Jenner to develop anthrax and rabies vaccines.

This is likely the most important life-saving discovery in medical history as various vaccines since then have prevented suffering and death for hundreds of millions of people. It is an ideal example of learning from limited explainable knowledge since little was known then beyond the observation that cowpox appeared to prevent smallpox.

It exemplifies the *Theory of Nothing* point stated in Chapter 4: we often do not need to explain everything about a thing to get useful results!

Autism

Autism is a disorder that affects social interaction and communication. It has proven extremely frustrating to parents with autistic children. Even though the genetic details have not been determined, all scientific indications are that autism has a genetic cause. Despite this, autism is a recent example of medicine being driven by frustration and hope in a wrong direction.

One study published in the medical journal *Lancet* in England in 1998 claimed a connection between autism and MMR (measles, mumps, rubella) vaccine. Many parents and autism groups were excited to have found an outside-of-the-body cause for autism. However, this study has been thoroughly discredited for scientific and ethical reasons. Indeed, in 2004, 10 of the 12 co-authors retracted the published interpretation of the study.

One unfortunate result is that much energy and work has been wasted by parents and autism groups. Another is that vaccinations of children decreased throughout England and North America which has resulted in increased illness and mortality for children.

Autism is a very puzzling ailment that has frustrated parents so much that common sense and science were perverted in order to find a cause. Autism is almost certainly caused by genes developed through random or inter-related processes. It will be discussed further in Chapter11, Genetics.

Ulcers

For about 100 years, doctors believed that most stomach and intestine ulcers were caused by stress and spicy foods. It wasn't until 1982 that two doctors - Barry Marshall and Robin Warren - discovered a type of bacteria that lives and grows in this part of the body and causes 80-90% of peptic ulcers. This showed that a specific physical effect (bacteria) is the primary cause rather than a more general mental or emotional factor (stress) which may be a secondary or aggravating factor.

This is an example of close observation and an open mind going beyond the conventional theories. And winning the Nobel Prize!

As will be seen in later chapters, the trend of physical or biochemical causes - primarily genetics - replacing mental or emotional or environmental causes has been the major medical story of the last 20 years or so. The body is created from - and defined by - genes: focussing on other factors is a denial of the body's origin. This illustrates the last part of principle 2 of the *Theory of Nothing*: a denial of their origins - and shows that other factors are (usually) wrong and lead to failure.

Chapter 9

Alternative Medicine

Alternative medicine strives to fill the holes of conventional medicine.

Alternative medicine in modern times has been embraced and developed with many theories and practices to fill the voids left by conventional medicine being unable to explain and integrate life or the human body into one encompassing theory or practice.

Some people latch onto alternative medicine with a fervour which likely stems from their longing for the certainty of explanation of life and the delay of mortality.

Alternative medicines have arisen over the centuries through one main route:

> Find some thing that works - and then develop and expand it into a theory and practice covering a system or component.

A small and brief sampling of alternative medicine illustrates this process.

Aromatherapy

This alternative medicine of using essential or aromatic oils has a long history. The modern name, aromatherapy, stems from a French chemist in the 1920's who accidentally discovered that lavender oil appeared to offer pain relief and healing enhancement for burns to human skin.

Even though certain essential oils may have value in treating certain things such as burns and gangrene, this does not mean that there is an overall effect or explanation for all essential oils.

Flower Remedies

This is an alternative medicine developed by the English Physician Edward Bach in the 1930's. Again, even though certain flower material may have value in treating certain things, this does not mean that there is an overall effect or explanation for all flower material.

Homeopathy

This alternative medicine uses serial dilution of substances that may produce similar symptoms of illnesses in healthy people. These, often extremely dilute, solutions of a wide range of substances are used to treat a wide range of illnesses.

Homeopathy is an extreme expansion of conventional oral vaccination; it treats many more illnesses by using much less of many more substances. Different theories have been advanced to explain how homeopathy works.

Homeopathy may have gained initial credence by default - it caused no harm in contrast to early medicines. Then, it likely benefited from the success of smallpox vaccines. Once again, even though small quantities of certain substances may have value in treating certain things, this does not mean that there is an overall effect or explanation for small quantities or for other substances.

Summary

All the foregoing examples illustrate the generalization of a specific to produce an overall theory or practice. It may be very satisfying to build an entire theory around some thing to make certain things explainable.

However, realize that application of a theory is unlikely to be able to predict and change. And be aware of the subconscious urge to force-fit facts or data to support the theory.

The value of each alternative medicine - as for everything else - is best achieved within the limits described by the *Theory of Nothing,* and by using the specific things that work.

Chapter 10

Mental Health

Mind comes from body biochemistry.

Psychologists and psychiatrists have had such a high failure rate in curing or even treating mental health problems because of two main factors:

(1) the attempt to fit people's problems into a system, often their own modification, of some theory or school of thought, and

(2) the fact that genes rule almost everything to a major degree.

Similarly, counsellors attempting to assist or treat people in changing or modifying their behaviour face a low success rate due to the foregoing two factors.

The major improvements and successes in mental health have come almost totally from medications. Even this tends to be a function of trial and error based on frustration that nothing else works.

Now we know that almost everything in the human body - both physical and mental - has a genetic or biochemical cause. This includes the entire mental constitution including personality, character, abnormalities and illness. Hence, medications are the most direct biochemical tool to address these mental characteristics.

Medications are usually a rough tool because of the complexity of the human body which is still being unravelled - mostly by genetics. As we achieve greater understanding of how genes work and which genes affect which parts of the human body and mind, we shall be able to develop more refined and effective treatments or even cures.

The next chapter will show some of the amazing genetic findings that hold out the promise for understanding and treating mental health.

Chapter 11

Genetics

Genetics is the body's expression of diversity.

Genetics applies the *Theory of Nothing* to the body. There is some order and relationship between certain genes but mostly they are unrelated end products of many long and different evolutionary - or even revolutionary - paths.

Several decades ago there was a strong belief in *mind over matter*, more precisely mind over body. However, the high failure rates of treating mental and personality disorders by counselling, psychology and psychiatry without medication was troubling. Also troubling was the lack of strong correlation between many physical health problems and diet or environmental factors.

Over the last 20 years genetics has come to embody and illustrate the power of: *body over mind.*

Further, genetics offers the best example and hope regarding the physical and mental health of humans.

The most common and useful way to correlate genes to different human characteristics has been to group people together who have the characteristic of interest, analyse their genes, and look for genetic differences from people who do not have that characteristic.

A second technique has been to study and manipulate the genes of mice or rats which have many similar genes to humans. This is often done to follow up on correlations suggested by the grouping studies.

A brief sampling of results from the field of genetics - exploding with research and results and possibilities - will be given. Every week there are reports of more things tied to human genes or more knowledge about specific genes.

Body Characteristics

Eye Colour has been studied for a number of years to show that it is inherited - genetic - and there are several genes involved. Originally, all humans had brown eyes.

A recent study led by Hans Eiberg from the University of Copenhagen found the precise component of a gene responsible for influencing the main eye colour gene to produce **Blue Eyes**. In addition, the research indicated that this was a genetic mutation which began with one person and likely occurred about 6-10,000 years ago in the Black Sea region.

This appears to be an example of an instantaneous or revolutionary change consistent with the *Theory of Nothing*.

Sports success often runs in families. Studies on twins by Hermine Maes from Virginia Commonwealth University showed that 67-90% of a child's physical and aerobic ability are related to heredity. Top athletes have a very good set of genes for their specific sport. Body type decides success in many sports. This should not stop people from practising sports - but realize that genes explain why most people cannot achieve certain levels of performance despite extensive training and practice.

Obesity results when body fat builds up past certain levels due to more calories being eaten than used up over time. Early thoughts were that obesity was merely a behavioural problem that could be overcome by eating less and exercising more.

It is now recognized that genes play a major role, directly and indirectly. Some genes may affect metabolism while others may affect behaviour towards food. Various studies have found from 11 to 127 genes associated with obesity. It is no wonder that so many people cannot control their weight!

These genes may have offered protection from malnutrition, illness and starvation during times when food was less plentiful. However, in regions of the Earth where food is now plentiful, this results in obesity for many people.

Obesity is a major health hazard associated with such common diseases as heart disease, hypertension, and diabetes. It is vital for the health of many people that genetics identify the various genes affecting obesity and clarify their effects and contribution so that treatments may be developed.

Personal Characteristics

Intelligence has been widely recognized for a long time as being inherited or genetic. However, there has been little research that identifies any specific gene. This is likely due to two reasons:

- less need to know compared to urgent 'problem' areas
- many genes are likely involved in various ways.

Recent research led by Katherine Burdick and Anil Malhotra of The Feinstein Institute for Medical Research illustrates the foregoing two reasons. This work was studying schizophrenia and identified a gene which appears to be associated with intelligence - and the gene only explained about 3% of the schizophrenia. Further, this gene had been previously associated with schizophrenia.

Another recent study led by Avshalom Caspi of King's College London has identified a variant of a gene which determines whether children get an intelligence boost from breastfeeding. This variant is also related to the ability to benefit from two fatty acids in breast milk.

These two genetic studies show the complexity and wide variety of genes affecting intelligence.

Happiness has been a characteristic which some people appear to have much more than others. A recent study led by Alexander Weiss of the University of Edinburgh used data from 900 pairs of twins. Evidence was found for certain genes that result in individuals tending to exhibit happiness regardless of circumstances.

Gambling is a problem throughout the world. Why do some people gamble to the extreme and others not at all? Pak Sham of King's College London is leading a genetic study of pathological gamblers in Hong Kong.

Cocaine addiction is another problem throughout the world. Gerome Breen of King's College London is leading a genetic study of cocaine addiction in Brazil.

Alcoholism is another widespread addiction problem which appears to be much worse amongst some societal racial communities such as the Inuit of northern Canada.

Researchers, led by Kenneth Blum at the University of Texas Health Science Center in 1990, put forth the DRD2 gene as the first specific gene associated with alcoholism. Since then, other genes such as CHRM2, GABRG3 and CREB have been identified as being related to alcoholism.

Genetics as a major factor has been backed up by family, twin and adoption studies which have shown such things as children of alcoholics being four times more likely than other children to become alcoholics. Twins reared in adoptive homes had a dramatically higher incidence of alcoholism if their biological fathers were alcoholics - irregardless of alcoholism in their adoptive families.

The Inuit of Canada most likely have a dangerous gene cocktail which makes them extremely susceptible to alcoholism.

Genetic knowledge to further the understanding of these three characteristic addictions will hopefully lead to better treatment to alleviate the devastating effects on addicted persons, their families and society.

Physical Health

Salt is an essential mineral for the human body. However, it results in high blood pressure for 60-70% of Canadians over age 60 due to excessive quantities or salt sensitivity. Studies have found that salt sensitivity varies widely amongst countries: 20% in South Korea to 50% in Canada to

80% in Portugal.

Frans Leenen of the University of Ottawa Heart Institute is co-leader of a study to identify a 'salt gene' in humans. As standard in genetic studies, the approach is to analyse and compare genes from different groups of people to find a gene or genes common only to the group with the problem.

There is very likely such an association, especially since researchers have bred rats to make some very salt resistant and others very salt sensitive. Once the genes are determined, closer studies will look at how they work - and how to counteract or moderate the effects of the 'salt genes'.

Breast cancer is the major cause of cancer death for North American women. In 1994 two genes were identified correlating to susceptibility to breast cancer: BRCA1 on chromosome 17 and BRCA2 on chromosome 13. Mutations in either of these genes results in a very significant chance of developing breast cancer. Recent studies on a related yeast protein indicate that these two genes repair DNA. Mutations in BRCA1 and BRCA2 are thought to disable this function to allow cancerous growth.

A new study led by Termumi Kohwi-Shigematsu of Lawrence Berkeley National Laboratory has identified a gene SATB1 on chromosome 3 which regulates the behaviour of 1,000 genes or more found inside tumour cells. Experiments - done on mice infected with human metastatic breast cancer cells - showed that deactivating SATB1 almost eliminated the formation of cancer cells whereas over-activating SATB1 greatly increased formation. Since SATB1 has a vital function in activating the immune system, treatment must only deactivate it in tumours, not throughout the body.

AIDS is a debilitating and often deadly illness caused by the HIV - a virus that is spread easily. Millions of people around the world suffer and are dying from AIDS. Much work has been done looking for a vaccine or for other treatments.

Rafick-Pierre Sekaly of the Université de Montréal led a team which has discovered a cell mechanism to reverse cell death in defective HIV cells. This was found by comparing and analysing the genes of a group of people infected with HIV but showing no symptoms to those of a group infected and showing HIV symptoms. In other words, some people have a natural immunity or protection against the negative effects of the HIV; the approach is to find what it is and to emulate it by some technique such as a vaccine.

Stephen Barr of the University of Alberta has discovered a gene that blocks late-stage HIV from multiplying and spreading. The gene and its protein are part of the body's natural defence system. However, more work needs to be done to find out why this gene is not working in people with HIV and how to turn it back on with drugs or vaccines.

ARVC5 is a heart disorder that causes sudden death by heart stoppage due to irregularity without prior warning. It affects people as young as 20-40 years of age - and runs in families. Males are 3 times as likely to be affected.

Terry-Lynn Young of Memorial University led a 12-year study which looked at the genes of 15 large Newfoundland families with the defect to narrow it down to chromosome 3 and then to FEB9 - a single gene with only one difference among

400 genetic letters.

This may be the most dramatic genetic result found so far: a single tiny difference or defect in only one gene having such a quick and deadly effect!

Not everyone in the affected families carries the defective gene. More than 100 individuals who are carrying the defective gene have decided to have defibrillators implanted near the collarbone. When an irregular heartbeat is detected, an automatic electrical shock restores a healthy heartbeat - saving them from instant death!

Mental Health

Autism is a complex brain disorder that inhibits a person's ability to communicate and develop social relationships along with extreme behavioural problems. It is likely caused by the interaction of multiple genes and is aggravated by normal social interactions. It exhibits a strong gender bias: the ratio of boys to girls for autism with lower intelligence is about 4 to 1, and about 10 to 1 for autism with normal intelligence.

A team from University of California Los Angeles has isolated a region of one gene (chromosome 17) that contributes to autism only in boys. This helps explain the much higher incidence in boys.

A study led by Pat Levitt of the Vanderbilt Kennedy Center for Research on Human Development has identified a single gene mutation (MET) that doubles a child's likelihood of having autism. This gene is not just a brain gene but one that is important for immune function and gut repair, as well. The effects of this gene mutation likely were one major factor that helped misguide parents and researchers to blame MMR vaccine as outlined in Chapter 8.

Another autism gene (neurexin 1) on Chromosome 11 was more recently located by studying 1200 families with multiple cases of autism involving more than 120 scientists from more than 50 institutions across 19 countries.

A study from Johns Hopkins University School of Medicine - again using large numbers of autistic children and their parents - was able to link autism risk to a specific variation in a gene (CNTNAP2) on Chromosome 7 related to neurexin proteins and brain function. There was a 20% higher incidence of the variant gene being inherited from the mother than from the father.

Researchers at Massachusetts General Hospital found that a child's autism risk increased by 100 times when a small part of Chromosome 16 was missing or duplicated. No such mutation was found in children without autism. However, only 1% of the autistic children studied have this mutation - and the parents do not have it. Hence, this appears to be a mutation that is not inherited - another type of process that may be instantaneous or spontaneous and is consistent with the *Theory of Nothing*

The foregoing 5 studies involve 5 different genes - proving the complexity of autism. They show that the autism genes cause other changes to the body besides the ones usually associated with autism. And there are likely more genes that contribute since none of these has proven to be dominant. No wonder autism has proved so frustrating to researchers - as well as parents and teachers.

Depression affects up to 20% of people and is about twice as common in women than men, perhaps evening out after age 55.

Researchers at King's College London looked at a group of 847 people in New Zealand and focussed on those who had suffered several stressful experiences between the ages of 21 and 26. The study looked at the 5-HTT gene which helps to control serotonin which is a brain chemical affecting mood. It was found that 43% of people with the two short variants of the gene developed depression after stressful events compared to only 17% who had two long variants of the gene.

A study led by Daniel Weinberger of NIMH used MRI to scan the brain of people viewing images of angry or fearful faces. People with two short variants of the 5-HTT gene showed decreased brain regions critical for the extinction of negative emotions.

A study led by Per Svenningsson of Rockefeller University shows that gene p11 is related to serotonin in the brain and may play a key role in depression. People with depression were shown to have decreased levels of the p11 protein. To test the relation, the researchers engineered two strains of mice: one that produced increased p11 and one that produced no p11. Mice with increased p11 acted like mice on anti-depressants and mice with no p11 acted depressed.

Manic depression or **bipolar** disorder is a mood disorder characterized by episodes of manic mood and behaviour often preceded or followed by episodes of depressed mood. It runs in families and studies on twins support the importance of genes. It may affect 1 in 100 people.

A team of British and Canadian researchers studied identical twins. They found that if one twin has manic depression, then the other has 85-89% likelihood of having it.

Columbia University is studying generations of families with manic depression to find the genes that cause this disorder. They have reported that chromosome 21 may contain one such gene. Other research groups have reported genes on chromosomes 4, 13, 18 and X as possible contributors. It appears to require a combination of genes to produce manic depression.

Mice are being used to insert, knock out or mutate mouse genes similar to humans to see how it affects behaviour of the changed mice. This is a widespread and powerful tool to follow up gene correlations found in groupings of humans.

Schizophrenia is a complex brain disorder affecting thoughts and perceptions which affects 1% of the world population. Evidence of the strong genetic influence comes from family, twin and adoption studies which suggest at least 80% genetic contribution.

German researchers at Julius-Maximilians University studied one large family with high incidence of catatonic schizophrenia. A mutation in gene WKL1 on chromosome 22 was found in 7 family members with the disease but not in 6 others with no schizophrenia symptoms. This gene produces a protein which may transmit electric currents along nerves.

Nobel laureate Susumu Tonegawa of MIT led a research team that studied the DNA from families showing schizophrenia. They found a defective version of gene PPP3CC on chromosome 8 that is related to production of calcineurin.

Genetically modified mice that did not produce calcineurin exhibited a wide variety of behaviour that mirrors human schizophrenia. Perhaps 20-30 genes are related to schizophrenia with many of them possibly related to calcineurin. Development of drugs to stimulate production of calcineurin offers treatment possibilities.

Researchers at the NIMH studied humans regarding the GRM3 gene on chromosome 7 which regulates glutamate involved in brain cells. GRM3 has two variants, one of which is more common in schizophrenics.

Research led by Philip Seeman of the University of Toronto indicate that the schizophrenic characteristic of being supersensitive to dopamine can be caused by mutations to many - rather than 2 or 3 - genes. They used mice to show that the brain may compensate for various gene mutations by becoming super-sensitive to dopamine. This suggests that looking for the dopamine supersensitive mechanism may be more fruitful than tracking down all the genes.

Researchers from Uppsala University studied families in northern Sweden to show a strong correlation between gene QKI on chromosome 6 and schizophrenia. Mutation of QKI is found in schizophrenics along with decreased myelin production which results in changed impulse transmission by nerve fibers. One drug company is pursuing this area for treatment.

A team from Edinburgh University studied 200 young people aged 16-25 who had at least two schizophrenic relatives. A variation of the neuregulin gene on chromosome 8 - affecting the nervous system - was strongly correlated to schizophrenic symptoms. Abnormal brain activity on brain scans was more likely for those with the variant gene.

The complexity of the cause of schizophrenia is illustrated by the foregoing small sampling which identifies 5 different genes on 4 different chromosomes. These genes deal with the nervous system, calcineurin, glutamate, and dopamine. And it is thought that there are many more genes involved.

Summary

So far, the best example of one gene being involved is for **ARVC5** above. This offers identification which is very useful to help decide on implantation of defibrillators. However, there still remains much work to be done to determine how to counteract or change the defective gene without causing other problems - which would offer a more elegant and direct solution.

Genetics offers the primary hope for improving the physical and mental health of humans. However, treatments will still be crude and difficult for most health problems because of the usual involvement of many genes - with each gene serving multiple functions.

There is much genetic work being done to develop vital knowledge and technologies to manipulate or counteract individual genes that have been identified as causing problems - but without causing other problems!

Unfortunately, the *Theory of Nothing* predicts no underlying system or order for the human body or for genes. Understanding individual genes and some of their inter-relations is the only productive approach.

Chapter 12

Group Behaviour

Groups are collections of unexplainable individuals.

There has been much interest in making human group behaviour explainable. How and why does an aggregate of individuals act as they do in various groups and activities? Much effort has been expended trying to predict and change group behaviour. Why is this so difficult?

Each person or group of persons constitutes divergent laws; they are each laws unto themselves - although with many commonalities and interactions - reflecting the chaos and disorder of the Universe. The *Theory of Nothing* comes into play yet again and prevents group behaviour from being explainable.

We shall briefly examine some areas of group behaviour to illustrate that only pieces are explainable.

Politics & Voting

The study of voting influences - how and why people vote - is a highly active field which is very imprecise and unexplainable. Election advertising is both planned and done on-the-fly to try to cover many possible ways and issues to influence people because no one knows what really works.

Predictions from polls - although often very accurate - are not true predictions since people are asked how they are going to vote.

Economics

Economics is very dependent upon complex interactions of many humans interacting with various levels of various financial systems through many influences, most of which are changing. How can we understand economics when we cannot explain the human mind?

The fact that human behaviour is changing is a result of principle 4 of the *Theory of Nothing* that everything in the Universe inherently incorporates changeability - which is the second reason why economics is mostly unexplainable.

We may be able to explain the effects of certain economic measures over a short period of time. The adjustment of interest rates by government is done because some short-term effects are explainable; however, it would not be necessary so frequently if economics were explainable.

Education

How should we teach? How does someone learn? There is no one way to teach everyone because of the complexities and differences in minds and bodies. Again, the fruitful approach is to realize this and look for what works for various types of people.

Crime

Why do people commit crimes? How do we prevent crime? This is another area that is unexplainable for the similar reasons above that human behaviour is unexplainable and, therefore, unpredictable. It is fraught with frustration and failure as evidenced by some countries having much higher crime or incarceration rates than others.

Sports

Sports predictions are usually exercises in frustration. Somebody is usually right only because there is a limited number of known results - and someone is bound to choose the correct one.

Team sports are pretty much unexplainable because they contain two collections of individuals. Daily proof of this is found in the gambling facilities around the world where almost no one wins over the long term - except the ‘house’.

Individual sports are much more explainable because they are much easier to analyse, predict and change since they involve only 1 person against some comparison such as time or another person.

Summary

Only small segments of some group behaviour over short periods of time are explainable - as outlined above - because of the complexities of individuals and groups who are governed by the *Theory of Nothing*.

Chapter 13

Nothing Is Plenty

The beauty of nothing is in everything.

We should accept the "unknowability" of everything - *Life is Unexplainable* - while we revel in this beautiful chaos of life on Earth - and in the Universe.

Rejoice in the human spirit for itself - and protect it with human rights.

To understand why we cannot understand everything is much better than being able to explain everything since we could not grasp everything anyway - and there would be nothing left for us to discover!

Should we now stop trying to learn and explain things?

No! But we can now realize limitations which will help us to avoid frustration and to focus on determining which components of life are the most useful or amenable to explanation.

We can continue to explore the unknown with the knowledge that there is still an infinite variety of components to study.

Nothing is Everything!

Bio

Eric Scheuneman was born in Ottawa, Canada where he began school before moving to Trenton, Ontario. He spent his formative teen years in Alberta, mostly Calgary. Eric benefited from the excellent Canadian public education system where he loved learning, sports, chess, dancing, and people - despite a severe stutter since Grade 2-3.

After graduating from Carleton University in Ottawa, he did his graduate work at the University of Calgary in theoretical and analytical organic chemistry. He joined the Federal Government of Canada as an environmental chemist.

After several years he changed fields and worked in energy conservation for the Federal Government where he wrote his first book, *Insulating Homes for Energy Conservation*, which became the official standard for Canada.

After becoming disabled and unable to work because of Chronic Fatigue Syndrome, he became a human rights fighter for disabled and other rights in Canada. He won two precedent-setting cases against the Federal Government and one against the local school board.

He has been very active as a father and as a member of his community of Perth, Ontario in education, the arts and sports - and dancing.

Despite two serious disabilities, he believes he has been very fortunate to have travelled the unexplainable and genetic paths governing his life - helped immensely by his ‘happy’ genes!

www.ingramcontent.com/pod-product-compliance
Ingram Content Group UK Ltd.
Pitfield, Milton Keynes, MK11 3LW, UK
UKHW040557210726
13854UKWH00007B/1220

9 781435 714380